S0-DFI-159

Caring for Your Pet Dinosaur

Taking Care of Your STEGOSAURUS

Gail Terp

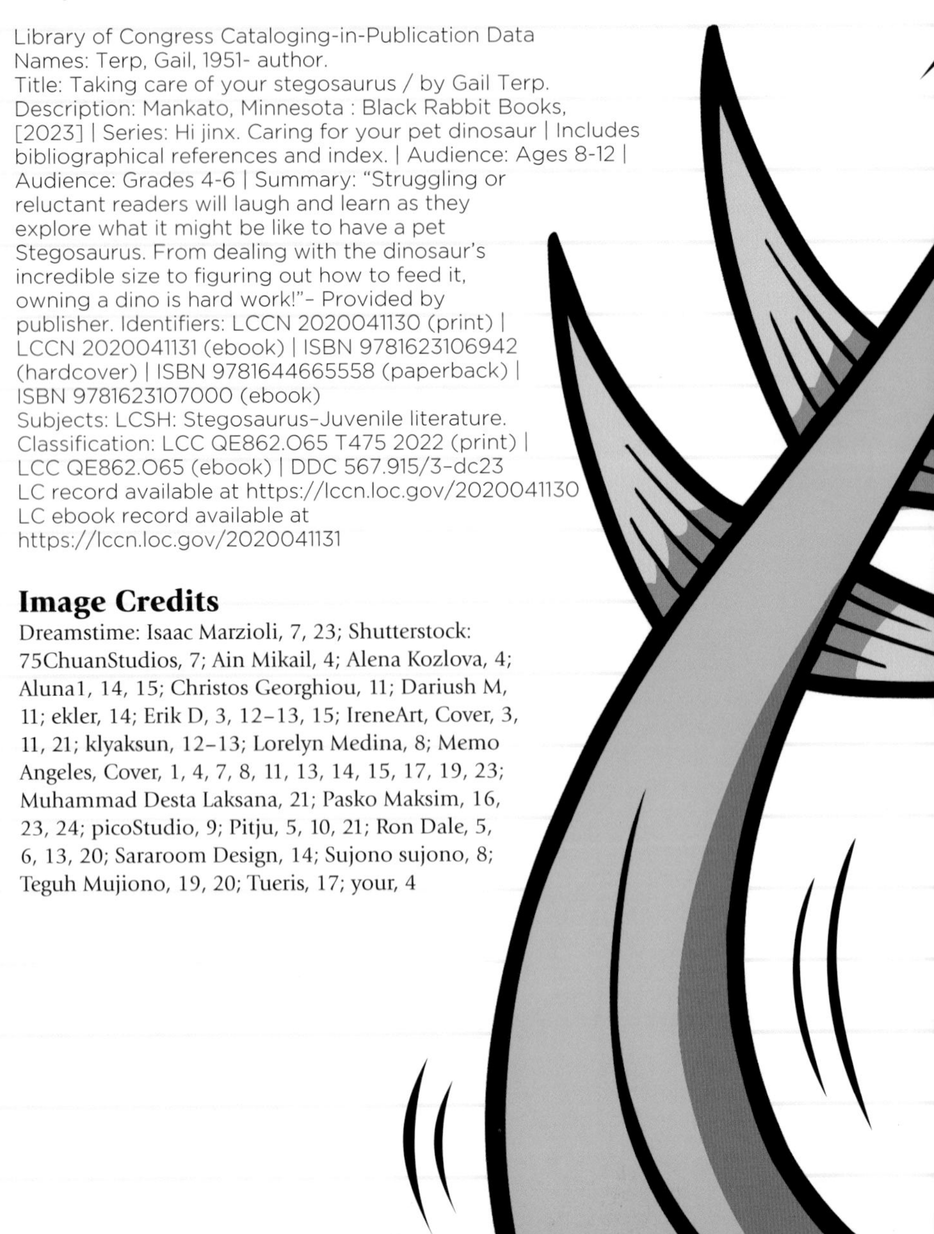

Hi Jinx is published by Black Rabbit Books
P.O. Box 227, Mankato, Minnesota, 56002.
www.blackrabbitbooks.com
Copyright © 2023 Black Rabbit Books

Gina Kammer & Marysa Storm editors;
Michael Sellner, designer and photo researcher

All rights reserved. No part of this book may be reproduced in any form without written permission from the publisher.

Library of Congress Cataloging-in-Publication Data
Names: Terp, Gail, 1951- author.
Title: Taking care of your stegosaurus / by Gail Terp.
Description: Mankato, Minnesota : Black Rabbit Books, [2023] | Series: Hi jinx. Caring for your pet dinosaur | Includes bibliographical references and index. | Audience: Ages 8-12 | Audience: Grades 4-6 | Summary: "Struggling or reluctant readers will laugh and learn as they explore what it might be like to have a pet Stegosaurus. From dealing with the dinosaur's incredible size to figuring out how to feed it, owning a dino is hard work!"- Provided by publisher. Identifiers: LCCN 2020041130 (print) | LCCN 2020041131 (ebook) | ISBN 9781623106942 (hardcover) | ISBN 9781644665558 (paperback) | ISBN 9781623107000 (ebook)
Subjects: LCSH: Stegosaurus-Juvenile literature.
Classification: LCC QE862.O65 T475 2022 (print) | LCC QE862.O65 (ebook) | DDC 567.915/3-dc23
LC record available at https://lccn.loc.gov/2020041130
LC ebook record available at
https://lccn.loc.gov/2020041131

Image Credits

Dreamstime: Isaac Marzioli, 7, 23; Shutterstock: 75ChuanStudios, 7; Ain Mikail, 4; Alena Kozlova, 4; Aluna1, 14, 15; Christos Georghiou, 11; Dariush M, 11; ekler, 14; Erik D, 3, 12–13, 15; IreneArt, Cover, 3, 11, 21; klyaksun, 12–13; Lorelyn Medina, 8; Memo Angeles, Cover, 1, 4, 7, 8, 11, 13, 14, 15, 17, 19, 23; Muhammad Desta Laksana, 21; Pasko Maksim, 16, 23, 24; picoStudio, 9; Pitju, 5, 10, 21; Ron Dale, 5, 6, 13, 20; Sararoom Design, 14; Sujono sujono, 8; Teguh Mujiono, 19, 20; Tueris, 17; your, 4

CONTENTS

Chapter 1

Is a STEGOSAURUS Right for You?

Looking for a pet but don't want a boring ol' cat or dog? Then a Stegosaurus might be right for you. This dino makes a big, fun-loving pet. Its giant plates are just one of its many **unique** features. But owning a Stegosaurus can be tricky. It can be expensive to feed. Its strange shape and size can cause problems. Think you're up to the task? Read on to find out.

Chapter 2

Understanding Your STEGOSAURUS

Adopting a Stegosaurus is a big step. It's important to know what you're getting into. You need to understand a Steggie's likes and dislikes. For example, a Steggie loves to have fun and laugh. Share dinosaur jokes with it. Tickle its toes. Make sure you have enough time to play with it. Your Steggie will feel lonely and left out if you don't.

happy Steggie

sad Steggie

Sightseeing Steggie

Stegosaurus also enjoys traveling. Take it sightseeing and to national parks. Be sure to pick up **souvenirs** wherever you go. Your Steggie will love having reminders of your adventures. Also be sure to keep your pet on a leash. If the dino gets excited and runs off, it could cause damage! The four 18-inch- (46-centimeter-) long spikes on its tail could take down trees.

People have discovered Stegosaurus fossils in Europe. Take your Steggie there to meet its cousins.

Fearful

You should know that this dino is afraid of some odd things. For example, cats scare it. So does the full moon in a starry night sky. But a Steggie's biggest fear isn't odd at all. It's absolutely terrified of big meat-eating dinosaurs. Make sure none of your neighbors have a pet Allosaurus. It'll probably try to eat your Steggie.

Allosaurus
full moon
cats

Chapter 3

Caring for Your STEGOSAURUS

Steggie care can be tough. To start, it needs a safe place to sleep. You'll have to build a barn big enough for it. This dino will grow up to 30 feet (9 meters) long. The barn must also be **sturdy**. When nervous, a Steggie swings its tail back and forth. The walls won't last long if they're not strong.

Feeding

Feeding a Steggie is a lot of work. This dino is an **herbivore** that eats lots of leafy bushes. You'll have to give it huge **bales** of leaves each day. Fruits, such as blueberries and strawberries, make special treats. But don't give it raspberries. It hates when raspberry seeds get stuck in its peglike teeth.

You'll notice that your Stegosaurus gets big cheeks when it eats. Don't worry! It'll swallow the food stored there later.

Grooming

This dino enjoys having its teeth brushed. You'll need a long-handled toothbrush to reach them all. Remember that Steggie likes mint toothpaste best.

Your pet will need weekly baths. When it's bath time, take your pet to the beach. It'll happily splash around to get clean. Use another big brush to help clean its plates. These plates are about 2 feet (1 m) tall.

Don't climb on your Steggie's plates. They're only attached to the skin. Climbing on them would hurt the dino.

A Wonderful Friend

Steggie makes a great pet for the right person. Do you have enough room? Do you tell good jokes? When the dino gets scared, can you **dodge** its tail? If so, a Steggie might be right for you.

Stegosaurus went **extinct** millions of years ago. Scientists can still learn about it by studying its fossils, though. These studies have given scientists ideas about the plates' purposes. The plates might have helped Stegosaurus protect itself from attackers. They could have helped keep it cool. The plates might have also helped attract **mates**. Scientists aren't sure which idea is correct. Maybe someday you'll be the one to find out!

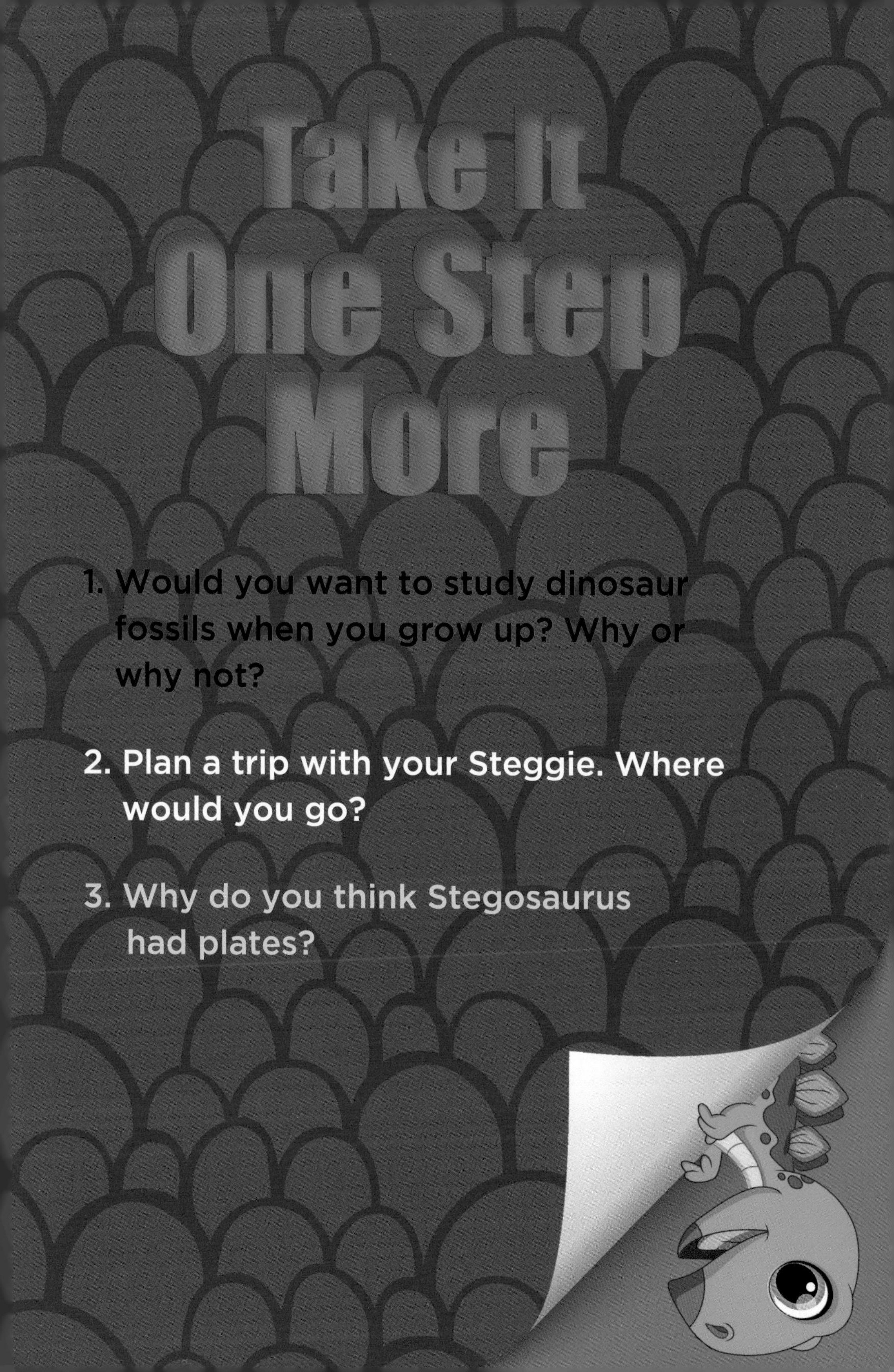

Take It One Step More

1. Would you want to study dinosaur fossils when you grow up? Why or why not?

2. Plan a trip with your Steggie. Where would you go?

3. Why do you think Stegosaurus had plates?

GLOSSARY

bale (BEYL)—a large bundle of goods tightly tied for storing or shipping

dodge (DOJ)—to move quickly in order to avoid being hit

extinct (ek-STINGKT)—no longer existing

fossil (FAH-sul)—the remains or traces of plants and animals that are preserved as rock

herbivore (HERB-uh-vor)—a plant-eating animal

mate (MAYT)—one of a pair that joins together to produce young

souvenir (SOO-veh-neer)—something kept as a reminder

sturdy (STUR-dee)—strong and healthy

unique (yoo-NEEK)—very special or unusual

LEARN MORE

BOOKS

Finn, Peter. *The Mighty Stegosaurus.* Dinosaur World. New York: Enslow Publishing, 2022.

Radley, Gail. *Stegosaurus.* Dinosaur Discovery. Mankato, MN: Black Rabbit Books, 2021.

Sabelko, Rebecca. *Stegosaurus.* The World of Dinosaurs. Minneapolis: Bellwether Media, Inc., 2020.

WEBSITES

Animals: Stegosaurus Dinosaur
www.ducksters.com/animals/stegosaurus.php

Stegosaurus
kids.nationalgeographic.com/animals/prehistoric-animals/stegosaurus/

Stegosaurus
www.dkfindout.com/us/dinosaurs-and-prehistoric-life/dinosaurs/stegosaurus/

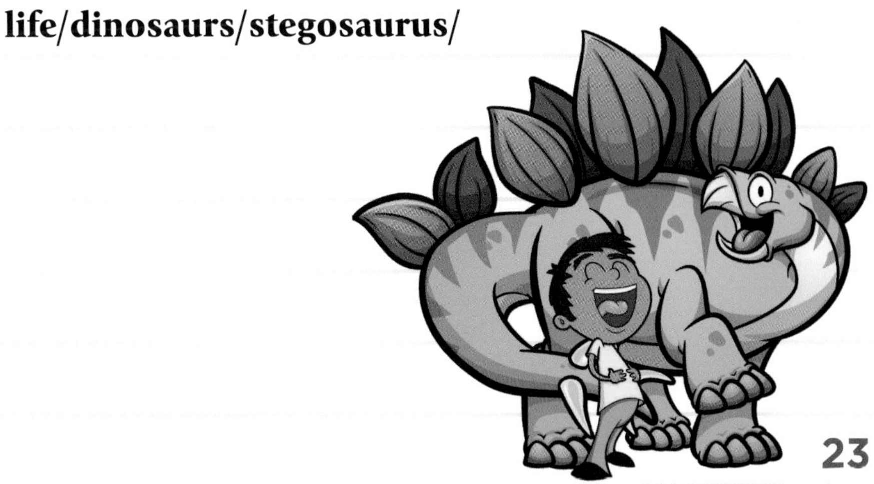

INDEX